FMEA Reference Guide

FMEA Investigator

Computer-Based Training Program

Interactive Learning Systems

Resource Engineering, Inc.
PO Box 219
Tolland, CT 06084
www.reseng.com
www.qualitytrainingportal.com

The *FMEA Reference Guide* is available for purchase from Resource Engineering or through any distributor of the *FMEA Investigator* computer-based training program. Special discount rates are given to users of the *FMEA Reference Guide* when the guides are purchased in bulk. For more information contact your distributor or:

Resource Engineering, Inc.
P.O. Box 219
Tolland, CT   06084-0219; USA
Phone: 800-810-8326 (North America only) or 860-872-2100
Fax: 860-872-2488
e-mail: sales@reseng.com

Second Edition — 2002
(In the prior edition, the *FMEA Reference Guide* was titled *FMEA Investigator Workbook*.)
Copyright © 1998 - 2002 Resource Engineering, Inc.

The Clip Art used in *FMEA Investigator* is from *Task Force Clip Art Really Big Edition*, New Vision Technologies, Inc. and *Corel Draw Clip Art*, Corel Corporation. All Rights Reserved.

Windows™, Word™, and Excel™ are trademarks of Microsoft® Corporation.

All other brand and product names are trademarks or registered trademarks of their respective companies.

Printed in the United States of America

07 06 05 04 03 02   10 9 8 7 6 5 4 3

Resource Engineering, Inc.
P.O. Box 219
384G Merrow Road
Tolland, CT   06084-0129
www.reseng.com
www.qualitytrainingportal.com

ISBN 1-882307-24-0

Contents

Welcome

Although Failure Mode and Effects Analysis (FMEA) techniques have been around for over 30 years, only recently have they gained popularity outside of the safety arena. This new interest is in large part due to the U.S. automotive industry and specifically its QS-9000 and TS 16949 supplier requirements. Other major U.S. industries, including aerospace, medical products, and electronics, are also using FMEA techniques as part of their improvement strategy.

Unlike many quality improvement tools, FMEAs do not require complicated statistics. FMEA studies can yield significant savings for a company as well as reduce the potential liability of a process or product that does not perform as promised.

FMEAs do take time and resources. Because the foundation of FMEAs is the input of team members, several people are typically involved in these studies. Proper training enables team members to work efficiently and effectively through the FMEA process. Until now, this type of training was available only through seminars and workshops.

The *FMEA Investigator*, a self-directed computer-based training program, is changing the way people learn about FMEAs. Now you and your employees can get high quality FMEA training on demand. You decide when you need the training and schedule it at your convenience. With this training you will learn a step-by-step process for conducting design and process FMEAs.

Additional training topics include the relationship between FMEAs and quality standards such as ISO 9000, QS-9000, and TS 16949 guidelines for customizing the FMEA ranking scales, and how FMEAs can be linked to Control Plans.

The *FMEA Reference Guide* is the official companion guide to the *FMEA Investigator*. In addition to a recap of the main points in the *FMEA Investigator*, you will also find copies of the worksheet formats used to conduct an FMEA and examples of Custom FMEA Ranking Scales in the Appendices.

Some FMEA learners follow along in the reference guide as they are going through the training program, taking notes as appropriate. Others use the book as a review tool or reference once their training is complete. Everyone, however, will find the *FMEA Reference Guide* an invaluable reference while conducting FMEAs and during FMEA team meetings.

Planning Your Training

The *FMEA Investigator* consists of three units: FMEA Overview, Design-FMEA, and Process-FMEA.

A key benefit of this training is that you can learn exactly what you need to know, when you need to learn it. Because your learning needs are unique, we suggest you customize your training by working through the units and lessons that will help you achieve your learning objectives. The table below provides some ideas.

What You Want to Learn	Experience with FMEA	Suggested Path
What an FMEA is & why are they used.	None.	▪ Unit 1: FMEA Overview, Lessons 1, 2, & 3
In general, how to conduct an FMEA.	None.	▪ Unit 1: FMEA Overview, Lessons 4, 5, & 6
How to conduct a Design-FMEA.	None.	▪ Unit 1: FMEA Overview ▪ Unit 2: Design-FMEAs
	Understand basic FMEA concepts.	▪ Unit 2: Design-FMEAs
How to conduct a Process-FMEA.	None.	▪ Unit 1: FMEA Overview ▪ Unit 3: Process-FMEAs
	Understand basic FMEA concepts.	▪ Unit 3: Process-FMEAs
Examples of how companies use FMEAs.	None.	▪ Unit 2, Lesson 5 (DFMEA) ▪ Unit 3, Lesson 5 (PFMEA)

Unit 1: FMEA Overview

Lesson 1 **Introduction**	• An overview of what an FMEA is; how the FMEA process works; and why an FMEA is used.
Lesson 2 **Purpose of an** **FMEA**	• An explanation of how an FMEA helps identify risks, prioritizes the risks relative to one another, and focuses efforts on an action plan to reduce the risks.
Lesson 3 **Tie to Quality** **Standards**	• An overview of the links between FMEAs and Quality Standards such as ISO 9000, QS-9000, & TS 16949.
Lesson 4 **DFMEA or** **PFMEA?**	• An explanation of the differences between a Design-FMEA and a Process-FMEA.
Lesson 5 **The FMEA** **Process**	• A preview of the 10 steps used to conduct an FMEA. The same basic steps apply to both a DFMEA and a PFMEA.
Lesson 6 **Assembling an** **FMEA Team**	• Helpful hints on assembling an effective FMEA team.
Investigative **Challenge**	• An assessment of the learner's progress in this unit.

Unit 2: Design-FMEAs

Lesson 1 **Design-FMEA** **Scope**	• How to clarify the scope for a DFMEA. • Details on how to use the DFMEA Scope Worksheet.
Lesson 2 **10 Steps to** **Conduct a DFMEA**	• Step-by-step directions on conducting a DFMEA. • Guidance on the use of the FMEA Analysis Worksheet. • Techniques for customizing the Severity, Occurrence, and Detection Ranking Scales for a DFMEA.
Lesson 3 **DFMEAs &** **Control Plans**	• Using the DFMEA Analysis to develop input for a Process Control Plan.
Lesson 4 **Getting More Out** **of Your DFMEA**	• Tips on the best times in a product's life cycle to conduct a DFMEA. • Tips on how to use the results of an FMEA to trigger continuous improvement.
Lesson 5 **DFMEA Example**	• An example of the application of a DFMEA, working through all 10 steps.
Investigative **Challenge**	• An assessment of the learner's progress in this unit.

Unit 3: Process-FMEAs

Lesson 1 **Process-FMEA** **Scope**	• How to clarify the scope for a PFMEA. • Details on how to use the PFMEA Scope Worksheet.

Lesson 2 **10 Steps to** **Conduct a PFMEA**	• Step-by-step directions on conducting a PFMEA. • Guidance on the use of the FMEA Analysis Worksheet. • Techniques for customizing the Severity, Occurrence, and Detection Ranking Scales for a PFMEA.

Lesson 3 **PFMEAs &** **Control Plans**	• Using the PFMEA Analysis to develop a proactive Control Plan.

Lesson 4 **Getting More Out** **of Your PFMEA**	• Tips on the best times and places to conduct a PFMEA. • Tips on how to use the results of an FMEA to trigger continuous improvement.

Lesson 5 **PFMEA Example**	• An example of the application of a PFMEA, working through all 10 steps.

Investigative **Challenge**	• An assessment of the learner's progress in this unit.

Unit 1 Objectives
FMEA Overview

✎ In this unit you will gain a basic understanding of the purpose, format, and application of FMEAs. Upon completion of this unit you will be able to:

- Explain the purpose of conducting an FMEA.

- Describe the link between FMEAs and ISO 9000, QS-9000, TS 16949 and other quality standards.

- Determine when to use a Design-FMEA and when to use a Process-FMEA.

- Explain the methodology of the FMEA process.

- Assemble an FMEA team.

Unit 1, Lesson 1:
Introduction

✎   FMEA means Failure Mode and Effects Analysis.

- Every product or process has modes of failure.
- The effects represent the impact of the failures.

✎   An FMEA is a tool to:

- Identify the relative risks designed into a product or process.
- Initiate action to reduce those risks with the highest potential impact.
- Track the results of the action plan in terms of risk reduction.

Unit 1, Lesson 2:
Purpose of an FMEA

✎ FMEAs help us focus on and understand the impact of potential process or product risks.

✎ A systematic methodology is used to rate the risks relative to each other.

● An RPN, or Risk Priority Number, is calculated for each failure mode and its resulting effect(s).

✎ The RPN is a function of three factors: The **Severity** of the effect, the frequency of **Occurrence** of the cause of the failure, and the ability to **Detect** the failure or effect.

● The RPN = The Severity ranking X the Occurrence ranking X the Detection ranking.

● The RPN can range from a low of 1 to a high of 1,000.

✎ Develop an Action Plan to reduce risks with unacceptably high RPNs.

✎ Use FMEAs as the basis for Control Plans. Control Plans are a summary of proactive defect prevention and reactive detection techniques.

Unit 1, Lesson 3:
Tie to Quality Standards

✎   ISO 9000 is the foundation for QS-9000 and TS 16949 (the quality standards for the automotive industry).

✎   Documents to support ISO 9000, QS-9000, TS 16949, and other industry standards can be helpful when conducting FMEAs.

✎   FMEAs fall under element 4.2 of the ISO 9000 standard.

✎   QS-9000 explicitly requires the use of both design and process FMEAs.

✎   The Quality Planning section of QS-9000 states:

"The supplier shall establish and implement an advanced product quality planning process. The supplier should convene internal multi-disciplinary teams to prepare for production of new or changed products. These teams should use appropriate techniques identified on the **Advanced Product Quality and Control Plan** reference manual. Similar techniques that accomplish the intent are acceptable.

Teams actions should include:

- Development/finalization of special characteristics

- Development and review of FMEAs

- Establishment of actions to reduce the potential failure modes with high risk priority numbers

- Develop or review of Control Plans"

Unit 1, Lesson 4:
DFMEA or PFMEA

✎　Similar principles and steps are followed for both design and process FMEAs.

✎　**Design-FMEAs**

- The primary objective of a Design-FMEA is to uncover potential failures associated with the product that could cause:
  - ✓　Product malfunctions.
  - ✓　Shortened product life.
  - ✓　Safety hazards while using the product.
- Design-FMEAs should be used throughout the design process—from preliminary design until the product goes into production.

✎　**Process-FMEAs**

- Process-FMEAs uncover potential failures that can:
  - ✓　Impact product quality.
  - ✓　Reduce process reliability.
  - ✓　Cause customer dissatisfaction.
  - ✓　Create safety or environmental hazards.
- Ideally, Process-FMEAs should be conducted prior to start-up of a new process, but they can be conducted on existing processes as well.

Unit 1, Lesson 5:
The FMEA Process

✎   There are 10 steps to the FMEA process. Both DFMEAs and PFMEAs use the same basic 10 steps.

✎   Teams, not individuals, conduct FMEAs.

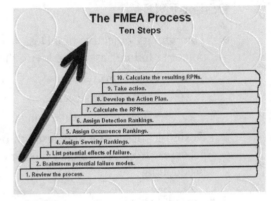

The FMEA Process
Ten Steps

10. Calculate the resulting RPNs.
9. Take action.
8. Develop the Action Plan.
7. Calculate the RPNs.
6. Assign Detection Rankings.
5. Assign Occurrence Rankings.
4. Assign Severity Rankings.
3. List potential effects of failure.
2. Brainstorm potential failure modes.
1. Review the process.

●   The skills and experience of a balanced team are needed to successfully complete an FMEA.

✎   The FMEA process involves the use of several forms including:

●   FMEA Team Start-Up Worksheet
●   FMEA Scope Worksheet (Design or Process)
●   FMEA Analysis Worksheet

✎   The FMEA process results in the assignment of risk priority numbers (RPNs) to each potential failure. Target failures with the highest RPNs for improvement.

✎   A Control Plan is a natural extension of an FMEA, even though it is not considered officially a part of an FMEA.

Unit 1, Lesson 6:
Assembling an FMEA Team

✎   FMEAs should always be conducted by teams.

✎   The best size for an FMEA team is 4 to 6 people, carefully selected, based on the contribution they can make to the specific FMEA.

✎   An FMEA team should represent a cross-section of the company in terms of functional responsibility and level in the organization. Team members typically come from:

- Manufacturing
- Materials/Purchasing
- Maintenance
- Tech Service
- Suppliers

- Engineering
- R & D
- Quality
- Customers

✎   FMEA team members do not necessarily need to have extensive knowledge of the design or process being targeted. In fact, sometimes it helps to get an outsider's fresh perspective.

✎   The FMEA team needs a leader to help set up and facilitate meetings, to ensure the team has the necessary resources, and to make sure the team is progressing toward completion of the FMEA.

Unit 1
Investigative Challenge

✎ Here are some things we suggest you do to prepare for the Investigative Challenge:

- Review the meaning of FMEA terms such as:
  - ✓ Severity, Occurrence, and Detection Ranking
  - ✓ RPN
  - ✓ Control Plan
- Understand what the concept of "relative risk" means and be able to describe how FMEAs are used to assess risk.
- Know how FMEAs and Control Plans are linked.
- Be able to describe how to assemble an FMEA team.
- Be able to explain the difference between Process-FMEAs and Design-FMEAs.

Unit 2 Objectives
Design-FMEAs

✎ In this unit you will learn everything you need to know to conduct Design-FMEAs. Upon completion of this unit you will be able to:

● Clarify the scope of a DFMEA.

● Work through the 10 steps of a DFMEA.

● Develop custom ranking scales for Severity, Occurrence, and Detection.

● Determine which technology tools to use as aids in your DFMEA action plan.

● Learn how to make the DFMEA into a living document.

● Link the DFMEA to a Control Plan.

Unit 2, Lesson 1:
Design-FMEA Scope

✎   Your DFMEA team will be most effective when the scope of the FMEA is well-defined.

✎   The DFMEA Scope Worksheet provides your team with the information necessary to clarify and fully understand the scope of the study.

✎   If the scope of a DFMEA seems too big, the FMEA team should consider breaking it up into two or three complementary studies.

DFMEA Scope Worksheet
Product————
Date————
Scope Defined by————

1. Who is the customer?
2. What are the product features and characteristics?
3. What are the product benefits?
4. Study the entire product or only components or sub-assemblies?
5. Include considerations of raw material failures?
7. Include packaging, storage, & transit?
8. What are the manufacturing process requirements & constraints?

✎   Once your FMEA team has defined the scope of the DFMEA, they should complete the FMEA Team Start-Up Worksheet. The worksheet will help clarify roles and responsibilities and define boundaries of freedom for the team.

FMEA Team Start-Up Worksheet
FMEA Number:———   Date Started ———
Team Members:———   Date Completed ———

Team Leader:———
Who will take minutes and maintain records?
1. What is the scope of the FMEA? Include a clear definition of the process (PFMEA) or product (DFMEA) to be studied. Attach the Scope Worksheet.

2. Are all affected areas represented?
     YES   NO   Action:———
3. Are different levels and types of knowledge represented on the teams?
     YES   NO   Action:———
4. Are customers or suppliers involved?
     YES   NO   Action:———
          FMEA Team Boundaries of Freedom
5. What aspects of the FMEA is the team responsible for?
     FMEA Analysis   Recommendations   Implementation of
                      for Improvement   Improvements
6. What is the budget for the FMEA? ———
7. Does the project have a deadline? ———
8. Do team members have specific time constraints? ———
9. What is the procedure if the team needs to expand beyond these boundaries?
10. How should the FMEA be communicated to others? ———

Unit 2, Lesson 2:
10 Steps to Conduct a DFMEA

Step 1: **Review the design**—Use a blueprint or schematic of the design/product to identify each component and interface.

Step 2: **Brainstorm potential failure modes**—Review existing documentation and data for clues.

Step 3: **List potential effects of failure**—There may be more than one for each failure.

Step 4: **Assign Severity rankings**—Based on the severity of the consequences of failure.

Step 5: **Assign Occurrence rankings**—Based on how frequently the cause of the failure is likely to occur.

Step 6: **Assign Detection rankings**—Based on the chances the failure will be detected prior to the customer finding it.

Step 7: **Calculate the RPN**—Severity x Occurrence x Detection.

Step 8: **Develop the action plan**—Define who will do what by when.

Step 9: **Take action**—Implement the improvements identified by your DFMEA team.

Step 10: **Calculate the resulting RPN** - Re-evaluate each of the potential failures once improvements have been made and determine their impact on the RPNs.

Step 1: Review the Design

✎ Reasons for the review:

- Help assure all team members are familiar with the product and its design.
- Identify each of the main components of the design and determine the function or functions of those components and interfaces between them.
- Make sure you are studying all components defined in the scope of the DFMEA.

✎ Use a print or schematic for the review.

  ● Add Reference Numbers to each component and interface.

✎ Try out a prototype or sample.

  ● Invite a subject matter expert to answer questions.
  ● Document the function(s) of each component and interface.

Step 2: Brainstorm Potential Failure Modes

✎ Consider potential failure modes for each component and interface.

  ● A potential failure mode represents any manner in which the product component could fail to perform its intended function or functions.
  ● Remember that many components will have more than one failure mode. Document each one. Do not leave out a potential failure mode because it rarely happens. Don't take shortcuts here; this is the time to be thorough.

✎ Prepare for the brainstorming activity.

  ● Before you begin the brainstorming session, review documentation for clues about potential failure modes.
  ● Use customer complaints, warranty reports, and reports that identify things that have gone wrong, such as hold tag reports, scrap, damage, and rework, as inputs for the brainstorming activity.
  ● Additionally, consider what may happen to the product under difficult usage conditions and how the product might fail when it interacts with other products.

Step 3: List Potential Effects of Failure

✎ The effect is related directly to the ability of that specific component to perform its intended function.

  ● An effect is the impact a failure could make should it occur.

- Some failures will have an effect on customers; others on the environment, the process the product will be made on, and even the product itself.

✎ The effect should be stated in terms meaningful to product performance. If the effects are defined in general terms, it will be difficult to identify (and reduce) true potential risks.

Step 4: Assign Severity Rankings

✎ The ranking scales are mission critical for the success of a DFMEA because they establish the basis for determining risk of one failure mode and effect relative to another.

✎ The same ranking scales for DFMEAs should be used consistently throughout an organization. This will make it possible to compare the RPNs from different FMEAs to one another.

- The severity ranking is based on a relative scale ranging from 1 to 10. A "10" means the effect has a dangerously high severity leading to a hazard without warning. Conversely, a severity ranking of "1" means the severity is extremely low. The scales provide a relative, not an absolute, scale.
- See Appendix 4 for the "Standard" (AIAG) DFMEA Severity Ranking Scale.

✎ The best way to customize a ranking scale is to start with a standard generic scale and then modify it to be more meaningful to your organization.

- By adding organization-specific examples to the ranking definitions, DFMEA teams will have an easier time using the scales. The use of examples saves teams time and improves the consistency of rankings from team to team.
- As you add examples specific to your organization, consider adding several columns with each column focused on a topic. One topic could provide descriptions of severity levels for customer satisfaction failures and another for environmental,

health, and safety issues. However, remember that each row should reflect the same relative impact, or severity, on the organization or customer.

● See Appendix 7 for examples of Custom DFMEA Ranking Scales. (Examples of custom scales for severity, occurrence, and detection rankings are included in this Appendix.)

Step 5: Assign Occurrence Rankings

✎ We need to know the potential cause to determine the occurrence ranking because, just like the severity ranking is driven by the effect, the occurrence ranking is a function of the cause.

● The occurrence ranking is based on the likelihood, or frequency, that the cause (or mechanism of failure) will occur.

● If we know the cause, we can better identify how frequently a specific mode of failure will occur.

✎ The occurrence ranking scale, like the severity ranking, is on a relative scale from 1 to 10.

● An occurrence ranking of "10" means the failure mode occurrence is very high; it happens all of the time. Conversely, a "1" means the probability of occurrence is remote.

● See Appendix 5 for the "Standard" (AIAG) DFMEA Occurrence Ranking Scale.

✎ Your organization may need to customize the occurrence ranking scale to apply to different levels or complexities of design. It is difficult to use the same scale for a modular design, a complex design, and a custom design.

● Some organizations develop three different occurrence ranking options (time-based, event-based, and piece-based) and select the option that applies to the design or product.

- See Appendix 7 for examples of Custom DFMEA Ranking Scales. (Examples of custom scales for severity, occurrence, and detection rankings are included in this Appendix.)

Step 6: Assign Detection Rankings

✎ To assign detection rankings, consider the design or product-related controls already in place for each failure mode and then assign a detection ranking to each control.

- Think of the detection ranking as an evaluation of the ability of the design controls to prevent or detect the mechanism of failure.

✎ Prevention controls are always preferred over detection controls.

- Prevention controls prevent the cause or mechanism of failure or the failure mode itself from occurring; they generally impact the frequency of occurrence. Prevention controls come in different forms and levels of effectiveness.

- Detection controls detect the cause, the mechanism of failure, or the failure mode itself after the failure has occurred BUT before the product is released from the design stage.

✎ A detection ranking of "1" means the chance of detecting a failure is almost certain. Conversely, a "10" means the detection of a failure or mechanism of failure is absolutely uncertain.

- See Appendix 6 for the "Standard" (AIAG) DFMEA Detection Ranking Scale.

- To provide DFMEA teams with meaningful examples of Design Controls, consider adding examples tied to the Detection Ranking scale for design related topics such as:
  ✓ Design Rules
  ✓ DFA/DFM (design for assembly and design for manufacturability) Issues
  ✓ Simulation and Verification Testing

● See Appendix 7 for examples of Custom DFMEA Ranking Scales. (Examples of custom scales for severity, occurrence, and detection rankings are included in this Appendix.)

Step 7: Calculate the RPN

✎ The RPN is the Risk Priority Number. The RPN gives us a relative risk ranking. The higher the RPN, the higher the potential risk.

✎ The RPN is calculated by multiplying the three rankings together. Multiply the Severity Ranking times the Occurrence Ranking times the Detection Ranking. Calculate the RPN for each failure mode and effect.

● **Editorial Note**: The current *FMEA Manual* from AIAG suggests only calculating the RPN for the highest effect ranking for each failure mode. We do not agree with this suggestion; we believe that if this suggestion is followed, it will be too easy to miss the need for further improvement on a specific failure mode.

✎ Since each of the three relative ranking scales ranges from 1 to 10, the RPN will always be between 1 and 1000. The higher the RPN, the higher the relative risk. The RPN gives us an excellent tool to prioritize focused improvement efforts.

Step 8: Develop the Action Plan

✎ Taking action means reducing the RPN. The RPN can be reduced by lowering any of the three rankings (severity, occurrence, or detection) individually or in combination with one another.

● A reduction in the Severity Ranking for a DFMEA is often the most difficult to attain. It usually requires a design change.

● Reduction in the Occurrence Ranking is accomplished by removing or controlling the potential causes or mechanisms of failure.

● And a reduction in the Detection Ranking is accomplished by adding or improving prevention or detection controls.

✎ What is considered an acceptable RPN? The answer to that question depends on the organization.

● For example, an organization may decide any RPN above a maximum target of 200 presents an unacceptable risk and must be reduced. If so, then an action plan identifying who will do what by when is needed.

✎ There are many tools to aid the DFMEA team in reducing the relative risk of those failure modes requiring action. The following recaps some of the most powerful action tools for DFMEAs.

● Design of Experiments (DOE)
  ✓ A family of powerful statistical improvement techniques that can identify the most critical variables in a design and the optimal settings for those variables.

● Mistake-Proofing (Poka Yoke)
  ✓ Techniques that can make it impossible for a mistake to occur, reducing the Occurrence ranking to 1.
  ✓ Especially important when the Severity ranking is 10.

● Design for Assembly and Design for Manufacturability (DFA/DFM)
  ✓ Techniques that help simplify assembly and manufacturing by modularizing product sub-assemblies, reducing components, and standardizing components.

● Simulations
  ✓ Simulation approaches include pre-production prototypes, computer models, accelerated life tests, and value-engineering analyses.

Step 9: Take Action

✎ The Action Plan outlines what steps are needed to implement the solution, who will do them, and when they will be completed.

✎ A simple solution will only need a Simple Action Plan while a complex solution needs more thorough planning and documentation.

- Most Action Plans identified during a DFMEA will be of the simple "who, what, & when" category. Responsibilities and target completion dates for specific actions to be taken are identified.

- Sometimes, the Action Plans can trigger a fairly large-scale project. If that happens, conventional project management tools such as PERT Charts and Gantt Charts will be needed to keep the Action Plan on track.

Step 10: Recalculate the Resulting RPN

✎ This step in a DFMEA confirms the action plan had the desired results by calculating the resulting RPN.

✎ To recalculate the RPN, reassess the severity, occurrence, and detection rankings for the failure modes after the action plan has been completed.

Unit 2, Lesson 3:
Linking DFMEAs & Control Plans

✐ A natural link exists between FMEAs and Control Plans.

✐ The purpose of a Control Plan is to:

● Minimize variation.

● Provide a structured approach for control.

● Establish a well thought-out reaction plan.

● Document and communicate failures and improvements.

● Enhance customer satisfaction.

✐ The primary reason for developing a Control Plan from a DFMEA is to use the DFMEA for setting process control procedures and identifying primary control points for the process the product will be made on.

● A Control Plan linked to the DFMEA provides a structured documentation and communication vehicle for the process control group.

✐ Key components of a Control Plan include the:

✓ The control factor.

✓ Specifications and tolerances of control factors.

✓ Measurement system or technique.

✓ Sample size and sample frequency.

✓ Control method.

✓ Reaction plan.

	Control Plan						
	Control Factor	Spec or Tolerance	Measure Technique	Sample Size	Sample Freq.	Control Method	Reaction Plan
F							
M							
E							
A							

Unit 2, Lesson 4:
Getting More out of DFMEAs

✎ DFMEAs should be conducted:

- On all new products.
    - ✓ DFMEAs should be conducted throughout the design cycle, beginning in the concept stage. Revise the DFMEA in the preliminary design stage, revise again in the prototype stage, and finalize the DFMEA in the final design stage.

- Whenever a change is to be made to a product.

- On existing products.
    - ✓ Use the Pareto Principle to decide which product or products to work on first.

✎ Make your FMEA a living document by adding to it and updating it whenever new information about the design or process develops.

✎ FMEAs should be updated when:

- Product or process improvements are made.
- New failure modes are identified.
- New data regarding effects are obtained.
- Root causes are determined.

Unit 2, Lesson 5:
Sample DFMEA

✎   This lesson takes you through a DFMEA conducted by a team at Bio-Plexus on the Punctur-Guard® Winged Set. Here is the checklist the team used to plan and complete their DFMEA:

- Complete the DFMEA Scope Worksheet.

- Complete the Team Start-Up Worksheet.

- Review a schematic of the winged set and add a reference number to each component and interface.

- Brainstorm ways that components and interfaces can fail in terms of quality, safety, and productivity from both the producer's and user's perspective.

- Determine the effects of the failure modes and assign a Severity ranking to each effect.

- Identify potential causes of the failure modes and assign an Occurrence ranking to each cause.

- List design controls and assign a Detection ranking.

- Calculate the RPN for each failure mode and effect.

- Determine which failure modes must be targeted for reduction and develop an action plan to address each item targeted.

- Execute the Action Plan.

- Assess the results of the Action Plan by reassigning Severity, Occurrence, and Detection rankings reflecting the improvements made. Recalculate the RPNs.

- Determine the overall impact of the DFMEA.

Unit 2
Investigative Challenge

✎ Here are some things we suggest you do to prepare for the Investigative Challenge:

- Know how to clarify the scope of a DFMEA.
- Be able to describe how:
    - ✓ The Severity ranking is related to the effect of a failure.
    - ✓ The Occurrence ranking is linked to the cause of a failure mode.
    - ✓ The Detection ranking is related to the ability of the design controls to prevent or to detect the mechanism of failure.
- Know how to calculate the RPN.
- Be comfortable with developing an Action Plan.
- Understand how to develop a Control Plan from a DFMEA.
- Know when to revisit DFMEAs to make them living documents.

Unit 3 Objectives
Process-FMEAs

In this unit you will learn everything you need to know to conduct Process-FMEAs. Upon completion of this unit you will be able to:

● Clarify the scope of a PFMEA.

● Work through the 10 steps of a PFMEA.

● Develop custom ranking scales for Severity, Occurrence, and Detection.

● Determine which technology tools to use as aids in your PFMEA action plan.

● Link the PFMEA to a Control Plan.

● Learn how to make the PFMEA into a living document.

Unit 3, Lesson 1:
Process-FMEA Scope

✎ Defining the scope for a PFMEA can be more difficult than for a DFMEA because a process often has more elements to cover than a design.

✎ Your PFMEA team will be most effective when the scope of the FMEA is well-defined.

✎ The PFMEA Scope Worksheet provides your team with the necessary information to clarify and fully understand the scope of the study.

✎ If the scope of a PFMEA seems too big, your team should consider breaking it up into two or three complementary studies.

✎ Once your FMEA team has defined the scope of the PFMEA, you should complete the FMEA Team Start-Up Worksheet. The worksheet will help clarify roles and responsibilities and define boundaries of freedom for the team.

Unit 3, Lesson 2:
10 Steps to Conduct a PFMEA

Step 1: **Review the process**—Use a process flowchart to identify each process component.

Step 2: **Brainstorm potential failure modes**—Review existing documentation and data for clues.

Step 3: **List potential effects of failure**—There may be more than one for each failure.

Step 4: **Assign Severity rankings**—Based on the severity of the consequences of failure.

Step 5: **Assign Occurrence rankings**—Based on how frequently the cause of the failure is likely to occur.

Step 6: **Assign Detection rankings**—Based on the chances the failure will be detected prior to the customer finding it.

Step 7: **Calculate the RPN**—Severity X Occurrence X Detection.

Step 8: **Develop the action plan**—Define who will do what by when.

Step 9: **Take action**—Implement the improvements identified by your PFMEA team.

Step 10: **Calculate the resulting RPN**—Re-evaluate each of the potential failures once improvements have been made and determine the impact of the improvements.

Step 1: Review the Process

✎ Review the process components and the intended function or functions of those components.

● Use of a detailed flowchart of the process or a traveler (or router) is a good starting point for reviewing the process.

✎ There are several reasons for reviewing the process:

- First, the review helps assure that all team members are familiar with the process. This is especially important if you have team members who do not work on the process on a daily basis.

- The second reason for reviewing the process is to identify each of the main components of the process and determine the function or functions of each of those components.

- Finally, this review step will help assure that you are studying all components of the process with the PFMEA.

✎ Using the process flowchart, label each component with a sequential reference number.

- These reference numbers will be used throughout the FMEA process.

- The marked-up flowchart will give you a powerful visual to refer to throughout the PFMEA.

✎ With the process flowchart in hand, the PFMEA team members should familiarize themselves with the process by physically walking through the process. This is the time to assure everyone on the team understands the basic process flow and the workings of the process components.

✎ For each component, list its intended function or functions.

- The function of the component is the value-adding role that component performs or provides.

- Many components have more than one function.

Step 2: Brainstorm Potential Failure Modes

✎ In Step 2, consider the potential failure modes for each component and its corresponding function.

● A potential failure mode represents any manner in which the component or process step could fail to perform its intended function or functions.

✎ Using the list of components and related functions generated in Step 1, as a team, brainstorm the potential failure modes for each function.

● Don't take shortcuts here; this is the time to be thorough.

✎ Prepare for the brainstorming session.

● Before you begin the brainstorming session, review documentation for clues about potential failure modes.

Step 3: List Potential Effects of Failure

✎ Determine the effects associated with each failure mode. The effect is related directly to the ability of that specific component to perform its intended function.

● An effect is the impact a failure could make if it occurred.
● Some failures will have an effect on the customers and others on the environment, the facility, and even the process itself.

✎ As with failure modes, use descriptive and detailed terms to define effects.

● The effect should be stated in terms meaningful to product or system performance.
● If the effects are defined in general terms, it will be difficult to identify (and reduce) true potential risks.

Step 4: Assign Severity Rankings

✎ Assign a severity ranking to each effect that has been identified.

● The severity ranking is an estimate of how serious an effect would be should it occur.

- To determine the severity, consider the impact the effect would have on the customer, on downstream operations, or on the employees operating the process.

✎ The severity ranking is based on a relative scale ranging from 1 to 10.

- A "10" means the effect has a dangerously high severity leading to a hazard without warning.
- Conversely, a severity ranking of "1" means the severity is extremely low.

✎ The ranking scales (for severity, occurrence, and detection) are mission critical for the success of a PFMEA because they establish the basis for determining risk of one failure mode and effect relative to another.

- The same ranking scales for PFMEAs should be used consistently throughout your organization. This will make it possible to compare the RPNs from different FMEAs to one another.
- See Appendix 8 for the "Standard" (AIAG) PFMEA Severity Ranking Scale.

✎ The best way to customize a ranking scale is to start with a standard, generic scale and then modify it to be more meaningful to your organization.

- As you add examples specific to your organization, consider adding several columns with each column focused on a topic.
- One topic could provide descriptions of severity levels for operational failures, another column for customer satisfaction failures, and a third for environmental, health, and safety issues.
- See Appendix 11 for examples of Custom PFMEA Ranking Scales. (Examples of custom scales for severity, occurrence, and detection rankings are included in this Appendix.)

Step 5: Assign Occurrence Rankings

✎ Next, consider the potential cause or failure mechanism for each failure mode; then assign an occurrence ranking to each of those causes or failure mechanisms.

✎ We need to know the potential cause to determine the occurrence ranking because, just like the severity ranking is driven by the effect, the occurrence ranking is a function of the cause. The occurrence ranking is based on the likelihood, or frequency, that the cause (or mechanism of failure) will occur.

✎ If we know the cause, we can better identify how frequently a specific mode of failure will occur. How do you find the root cause?

● There are many problem-finding and problem-solving methodologies.

● One of the easiest to use is the 5-Whys technique.

● Once the cause is known, capture data on the frequency of causes. Sources of data may be scrap and rework reports, customer complaints, and equipment maintenance records.

✎ The occurrence ranking scale, like the severity ranking, is on a relative scale from 1 to 10.

● An occurrence ranking of "10" means the failure mode occurrence is very high, and happens all of the time. Conversely, a "1" means the probability of occurrence is remote.

● See Appendix 9 for the "Standard" (AIAG) PFMEA Occurrence Ranking Scale.

✎ Your organization may need an occurrence ranking scale customized for a low-volume, complex assembly process or a mixture of high-volume, simple processes and low-volume, complex processes.

- Consider customized occurrence ranking scales based on time-based, event-based, or piece-based frequencies.
- See Appendix 11 for examples of Custom PFMEA Ranking Scales. (Examples of custom scales for severity, occurrence, and detection rankings are included in this Appendix.)

Step 6: Assign Detection Rankings

To assign detection rankings, identify the process or product related controls in place for each failure mode and then assign a detection ranking to each control. Detection rankings evaluate the current process controls in place.

- A control can relate to the failure mode itself, the cause (or mechanism) of failure, or the effects of a failure mode.
- To make evaluating controls even more complex, controls can either prevent a failure mode or cause from occurring or detect a failure mode, cause of failure, or effect of failure after it has occurred.
- Note that prevention controls cannot relate to an effect. If failures are prevented, an effect (of failure) cannot exist!

The Detection ranking scale, like the Severity and Occurrence scales, is on a relative scale from 1 to 10.

- A Detection ranking of "1" means the chance of detecting a failure is certain.
- Conversely, a "10" means there is absolute certainty of non-detection. This basically means that there are no controls in place to prevent or detect.
- See Appendix 10 for the "Standard" (AIAG) PFMEA Detection Ranking Scale.

Taking a lead from AIAG, consider three different forms of Custom Detection Ranking options. Custom examples for Mistake-Proofing, Gauging, and Manual Inspection controls can be helpful to PFMEA teams.

- See Appendix 11 for examples of Custom PFMEA Ranking Scales. (Examples of custom scales for severity, occurrence, and detection rankings are included in this Appendix.)

Step 7: Calculate the RPN

✎ The RPN is the Risk Priority Number. The RPN gives us a relative risk ranking. The higher the RPN, the higher the potential risk.

✎ The RPN is calculated by multiplying the three rankings together. Multiply the Severity ranking times the Occurrence ranking times the Detection ranking. Calculate the RPN for each failure mode and effect.

- **Editorial Note**: The current *FMEA Manual* from AIAG suggests only calculating the RPN for the highest effect ranking for each failure mode. We do not agree with this suggestion; we believe that if this suggestion is followed, it will be too easy to miss the need for further improvement on a specific failure mode.

✎ Since each of the three relative ranking scales ranges from 1 to 10, the RPN will always be between 1 and 1000. The higher the RPN, the higher the relative risk. The RPN gives us an excellent tool to prioritize focused improvement efforts.

Step 8: Develop the Action Plan

✎ Taking action means reducing the RPN. The RPN can be reduced by lowering any of the three rankings (severity, occurrence, or detection) individually or in combination with one another.

- A reduction in the Severity ranking for a PFMEA is often the most difficult. It usually requires a physical modification to the process equipment or layout.

● Reduction in the Occurrence ranking is accomplished by removing or controlling the potential causes.

✓ Mistake-proofing tools are often used to reduce the frequency of occurrence.

● A reduction in the Detection ranking can be accomplished by improving the process controls in place.

✓ Adding process fail-safe shut-downs, alarm signals (sensors or SPC), and validation practices including work instructions, set-up procedures, calibration programs, and preventative maintenance are all detection ranking improvement approaches.

✎ What is considered an acceptable RPN? The answer to that question depends on the organization.

● For example, an organization may decide any RPN above a maximum target of 200 presents an unacceptable risk and must be reduced. If so, then an action plan identifying who will do what by when is needed.

✎ There are many tools to aid the PFMEA team in reducing the relative risk of failure modes requiring action. Among the most powerful tools are Mistake-Proofing, Statistical Process Control, and Design of Experiments.

● Mistake-Proofing (Poka Yoke)

✓ Techniques that can make it impossible for a mistake to occur, reducing the Occurrence ranking to 1.

✓ Especially important when the Severity ranking is 10.

● Statistical Process Control (SPC)

✓ A statistical tool that helps define the output of a process to determine the capability of the process against the specification and then to maintain control of the process in the future.

● Design of Experiments (DOE)

✓ A family of powerful statistical improvement techniques that can identify the most critical variables in a process and the optimal settings for these variables.

Step 9: Take Action

✎   The Action Plan outlines what steps are needed to implement the solution, who will do them, and when they will be completed.

✎   A simple solution will only need a Simple Action Plan while a complex solution needs more thorough planning and documentation.

●   Most Action Plans identified during a PFMEA will be of the simple "who, what, & when" category. Responsibilities and target completion dates for specific actions to be taken are identified.

●   Sometimes, the Action Plans can trigger a fairly large-scale project. If that happens, conventional project management tools such as PERT Charts and Gantt Charts will be needed to keep the Action Plan on track.

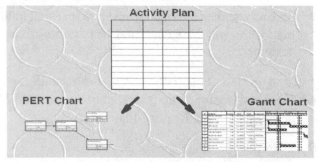

Step 10: Recalculate the Resulting RPN

✎   This step in a PFMEA confirms the action plan had the desired results by calculating the resulting RPN.

✎   To recalculate the RPN, reassess the severity, occurrence, and detection rankings for the failure modes after the action plan has been completed.

Unit 3, Lesson 3:
Linking PFMEAs & Control Plans

✎ Control Plans assure a system is in place to control the risks of the same failure modes as identified in the PFMEA. While Control Plans can be developed independently of PFMEAs, it is time and cost-effective to link Control Plans directly to PFMEAs.

✎ The primary intent of Control Plans is to create a structured approach for control of process and product characteristics while focusing the organization on characteristics important to the customer.

- A Control Plan does assure well thought-out reaction plans are in place in case an out-of-control condition occurs and provides a central vehicle for documentation and communication of control methods.
- Special attention is typically given to potential failures with high RPNs and those characteristics that are critical to the customer.

✎ A Control Plan deals with the same information explored in a FMEA plus more. The major additions to the FMEA needed to develop a Control Plan are:

- Identification of the control factors.
- The specifications and tolerances.
- The measurement system.
- Sample size.
- Sample frequency.
- The control method.
- The reaction plan.

✎ Don't let Control Plans become static.

● Just like work instructions, make Control Plans a living document.

● As changes in product or process characteristics, specifications, measurements systems, sampling, control methods, or the reaction plan are identified, update the control plan.

● Use the revision as a communication tool to spread the word of the changes to the supply chain.

● By making the FMEA a living document, you can be sure that potentials for failure are continually being eliminated or reduced.

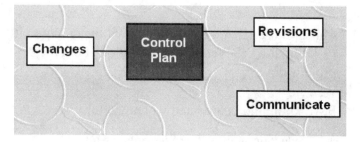

Unit 3, Lesson 4:
Getting More out of PFMEAs

✎ PFMEAs should be conducted:

- On all new processes.
  - ✓ PFMEAs should be conducted throughout the process design cycle beginning in the preliminary design stage.
  - ✓ Revise the PFMEA in the pilot process stage, revise again in the final design stage, and finalize the PFMEA in the as-built stage.
- Whenever a change is to be made to a process.
- On existing processes.
  - ✓ Use the Pareto Principle to decide which processes to work on first.

✎ Make your FMEA a living document by adding to it and updating it whenever new information about the design or process develops.

✎ FMEAs should be updated when:

- Product or process improvements are made.
- New failure modes are identified.
- New data regarding effects are obtained.
- Root causes are determined.

✎ Link FMEAs to Control Plans.

Unit 3, Lesson 5:
Sample PFMEA

✎ This lesson takes you through a PFMEA conducted by a team at Sermatech Klock on the brazing process for the nozzle seal used in the engine of the U.S. Army's M1A2 tank. Here is the checklist the team used to plan and complete their PFMEA:

- Complete the PFMEA Scope Worksheet.
- Complete the Team Start-Up Worksheet.
- Walk through the process and review the job traveler. Flowchart the major steps of the process and then add process substeps to the major steps.
- Brainstorm ways that the process can fail in terms of quality, safety, and productivity.
- Determine the effects of the failure modes and assign a Severity ranking to each effect.
- Identify potential causes of the failure modes and assign an Occurrence ranking to each cause.
- List detection controls and assign a Detection ranking for each control.
- Calculate the RPN for each failure mode and effect.
- Determine which failure modes must be targeted for reduction and develop an action plan to address each item targeted.
- Execute the Action Plan.
- Assess the results of the Action Plan by reassigning Severity, Occurrence, and Detection rankings reflecting the improvements made. Recalculate the RPNs.
- Determine the overall impact of the PFMEA.

Unit 3
Investigative Challenge

✎ Here are some things we suggest you do to prepare for the Investigative Challenge:

- Know how to clarify the scope of a PFMEA.
- Be able to describe how:
  - ✓ The Severity ranking is related to the effect of a failure.
  - ✓ The Occurrence ranking is linked to the cause of a failure mode.
  - ✓ The Detection ranking is connected to the process design.
- Know how to calculate the RPN.
- Be comfortable with developing an Action Plan.
- Understand how to develop a Control Plan from a PFMEA.

List of Appendices

The following 13 Appendices will be useful to FMEA teams and practitioners planning and conducting FMEAs.

1. **FMEA Team Start-Up Worksheet**

2. **Design-FMEA Scope Worksheet**

3. **Process-FMEA Scope Worksheet**

4. **Standard (AIAG) DFMEA Severity Ranking Scale**

5. **Standard (AIAG) DFMEA Occurrence Ranking Scale**

6. **Standard (AIAG) DFMEA Detection Ranking Scale**

7. **Examples of Custom DFMEA Ranking Scales (4 pages)**

8. **Standard (AIAG) PFMEA Severity Ranking Scale**

9. **Standard (AIAG) PFMEA Occurrence Ranking Scale**

10. **Standard (AIAG) PFMEA Detection Ranking Scale**

11. **Examples of Custom PFMEA Ranking Scales (5 pages)**

12. **FMEA Analysis Worksheet (2 pages)**

13. **Glossary of Terms**

Appendix 1
FMEA Team Start-Up Worksheet

FMEA Number: _____   Date Started: _____

Team   Date Completed: _____

Members: _____   _____   _____

_____   _____   _____

_____   _____   _____

Leader: _____

Who will take minutes and maintain records?   _____

---

1. What is the scope of the FMEA? Include a clear definition of the process (PFMEA) or product (DFMEA) to be studied. (Attach the Scope Worksheet.)

---

2. Are all affected areas represented? (circle one)

   YES   NO   Action: _____

3. Are different levels and types of knowledge represented on the team? (circle one)

   YES   NO   Action: _____

4. Are customer or suppliers involved? (circle one)

   YES   NO   Action: _____

Boundaries of Freedom

5. What aspect of the FMEA is the team responsible for? (circle one))

   FMEA Analysis   Recommendations for   Implementation of
                         Improvement   Improvements

6. What is the budget for the FMEA?   _____

7. Does the project have a deadline?   _____

8. Do team members have specific time
   constraints?   _____

9. What is the procedure if the team needs to
   expand beyond these boundaries?   _____

10. How should the FMEA be communicated to
    others?   _____

Appendix 2
Design FMEA Scope Worksheet

Product: _____

Date: _____

Scope Defined by: _____

Part 1: Who is the customer?

Part 2: What are the product features and characteristics?

Part 3: What are the product benefits?

Part 4: Study the entire product or only components or sub-assemblies?

Part 5: Include consideration of raw material failures?

Part 6: Include packaging, storage, & transit?

Part 7: What are the operational process requirements & constraints?

Appendix 3
Process FMEA Scope Worksheet

Process: _____

Date: _____

Scope Defined by: _____

Part 1: What process components are to be included in the investigation?

Part 2: Who is the customer?

Part 3: What process support systems are to be included in the study?

Part 4: To what extent should input materials be studied during the investigation?

Part 5: What are the product material requirements & constraints?

Part 6: Should packaging, storage, and transit be considered part of this study?

Appendix 4
Standard (AIAG) Design FMEA Severity Ranking Scale

(Should be tailored to meet the needs of your company.)

Ranking	Description	Definition (Severity of Effect)
10	Hazardous without warning	Very high severity ranking when a potential failure mode affects safe vehicle operation and/or involves noncompliance with federal regulations without warning.
9	Hazardous with warning	Very high severity ranking when a potential failure mode affects safe vehicle operation and/or involves noncompliance with federal regulations with warning.
8	Very high	Vehicle/item inoperable (loss of primary function).
7	High	Vehicle/item operable but at reduced level of performance. Customer very dissatisfied.
6	Moderate	Vehicle/item operable but Comfort/Convenience item(s) inoperable. Customer dissatisfied.
5	Low	Vehicle/item operable but Comfort/Convenience item(s) operable at a reduced level of performance. Customer somewhat dissatisfied.
4	Very Low	Fit & Finish/Squeak & Rattle item does not conform. Defect noticed by most customers (greater than 75%).
3	Minor	Fit & Finish/Squeak & Rattle item does not conform. Defect noticed by 50% of customers.
2	Very Minor	Fit & Finish/Squeak & Rattle item does not conform. Defect noticed by discriminating customers (less than 25%).
1	None	No discernible effect.

Reprinted with permission from the **FMEA Manual** (DaimlerChrysler, Ford Motor Company, General Motors Supplier Quality Requirements Task Force).

Appendix 5
Standard (AIAG) Design FMEA
Occurrence Ranking Scale

(Should be tailored to meet the needs of your company.)

Ranking	Description	Possible Failure Rate
10	**Very High:** Persistent failures.	≥ 100 per thousand vehicles/items.
9		50 per thousand vehicles/items.
8	**High:** Frequent failures.	20 per thousand vehicles/items.
7		10 per thousand vehicles/items.
6	**Moderate:** Occasional failures.	5 per thousand vehicles/items.
5		2 per thousand vehicles/items.
4		1 per thousand vehicles/items.
3	**Low:** Relatively few failures.	0.5 per thousand vehicles/items.
2		0.1 per thousand vehicles/items.
1	**Remote:** Failure is unlikely.	<0.010 per thousand vehicles/items.

Reprinted with permission from the **<u>FMEA Manual</u>** (DaimlerChrysler, Ford Motor Company, General Motors Supplier Quality Requirements Task Force).

Appendix 6
Standard (AIAG) Design FMEA Detection Ranking Scale

(Should be tailored to meet the needs of your company)

Rating	Detection	Criteria: Likelihood of Detection by Design Control
10	Absolute Uncertainty	Design Control will not and/or cannot detect a potential cause/mechanism and subsequent failure mode; or there is no Design Control.
9	Very Remote	Very remote chance the Design Control will detect a potential cause/mechanism and subsequent failure mode.
8	Remote	Remote chance the Design Control will detect a potential cause/mechanism and subsequent failure mode.
7	Very Low	Very low chance the Design Control will detect a potential cause/mechanism and subsequent failure mode.
6	Low	Low chance the Design Control will detect a potential cause/mechanism and subsequent failure mode.
5	Moderate	Moderate chance the Design Control will detect a potential cause/mechanism and subsequent failure mode.
4	Moderately High	Moderately high chance the Design Control will detect a potential cause/mechanism and subsequent failure mode.
3	High	High chance the Design Control will detect a potential cause/mechanism and subsequent failure mode.
2	Very High	Very high chance the Design Control will detect a potential cause/mechanism and subsequent failure mode.
1	Almost Certain	Design Control will almost certainly detect a potential cause/mechanism and subsequent failure mode.

Reprinted with permission from the **FMEA Manual** (DaimlerChrysler, Ford Motor Company, General Motors Supplier Quality Requirements Task Force).

Appendix 7 (Page 1 of 4)
Examples of a Custom DFMEA Scales

Severity: DFMEA Customer Satisfaction Examples

Ranking	Example
10	In-service failure that threatens safety.
9	Extensive product recall.
8	Unscheduled engine removal.
7	Premature (unscheduled) component replacement.
6	Oil leak but system still operational.
5	Air-conditioning system not operating properly.
4	Interior panel rattles.
3	Variation in seat colors.
2	Door plugs missing.
1	Scratch on interior of housing.

Severity: DFMEA EH&S Ranking Examples

Ranking	Example
10	Catastrophic product failure causes loss of life or serious injury.
9	Product creates major hazardous environmental disposal problem.
8	Use of product under normal conditions leads to OSHA Recordable injury.
7	Use of product under normal conditions leads to exposure above PEL.
6	Product creates moderate hazardous environmental disposal problem.
5	Manufacture of or use of product leads to temporary non-compliance with ISO 14001 audit.
4	Use of product under normal conditions leads to injury requiring first aid.
3	Use of product leads to spill of non-hazardous material.
2	Use of product leads to poor housekeeping.
1	Manufacture or use does not have a detectable impact on EH&S.

Appendix 7 (Page 2 of 4)

Occurrence: DFMEA Time-Based Ranking Examples

Ranking	Example
10	≥2 occurrences per day
9	≥1 occurrence per day
8	≥1 per 2-3 days
7	≥1 per week
6	≥1 per 2 weeks
5	≥1 per month
4	≥1 per quarter
3	≥1 per half-year
2	≥1 per year
1	<1 per 1 year

Occurrence: DFMEA Piece-Based Ranking Examples

Ranking	Example
10	Cpk < 0.33
9	Cpk ≈ 0.33
8	Cpk ≈ 0.67
7	Cpk ≈ 0.83
6	Cpk ≈ 1.00
5	Cpk ≈ 1.17
4	Cpk ≈ 1.33
3	Cpk ≈ 1.67
2	Cpk ≈ 2.00
1	Cpk > 2.00

Appendix 7 (Page 3 of 4)

Occurrence: DFMEA Event-Based Ranking Examples

Ranking	Example
10	≥5 per design
9	≥2
8	≥1
7	≥1:2 designs
6	≥1:5
5	≥1:10
4	≥1:50
3	≥1:100
2	≥1:250
1	<1:250

Detection: DFMEA Design Rules Ranking Examples

Ranking	Example
10	No design rules used.
9	Design protocols are formalized.
8	Design rules are specified in initial design criteria.
7	Design reviews held to ensure compliance to design rules.
6	Checklist used to ensure design rules are followed.
5	Purchasing systems do not allow selection of non-standard components.
4	Early supplier involvement so all relevant knowledge about input materials and compliance to design needs are understood.
3	Design software signals compliance issues.
2	Design software ensures compliance to the relevant industry standards.
1	Design software prevents use of non-standard dimensions, spacing, and tolerances.

Appendix 7 (Page 4 of 4)

Detection: DFMEA DFA/DFM Ranking Examples

Ranking	Example
10	No consideration given for DFA/DFM.
9	The number of components has been minimized.
8	Only standard components have been used.
7	Ergonomic assembly techniques have been incorporated.
6	Design elements such as pad sizes, wire gauge, and fasteners have been standardized throughout the design.
5	Modular designs used.
4	Easy-fastening devices (snap fits or quick fastening devices such as quarter-turn screw, twist locks, spring clips, latches) used.
3	Self-testing or self-diagnosis has been built-in.
2	Self-aligning surface, grooves, and guides used.
1	Asymmetrical features used to mistake-proof assembly.

Detection: Simulations & Verification Testing Examples

Ranking	Example
10	No verification testing used.
9	GO/NOGO tests used to ensure dimensional requirements.
8	Partial functionality of prototype tested before release.
7	Full Alpha tests conducted; no Beta testing.
6	Untested computer model used to simulate product performance.
5	Accelerated life testing of final design before release; Lab simulation.
4	Alpha and Beta testing used before release to ensure design meets needs.
3	Product tested for full functionality in customer's application.
2	Finite element analysis to highlight stress concentrations requiring design changes early in the design stages.
1	Computer modeling to ensure form and fit of mating components.

Appendix 8
Standard (AIAG) Process FMEA Severity Ranking Scale

(Should be tailored to meet the needs of your company.)

Effect	Customer Effect	Manufacturing Effect	Ranking
Hazardous without warning	Potential failure mode affects safe operation and/or involves noncompliance with regs without warning.	May endanger operator, machine, or assembly without warning.	10
Hazardous with warning	Potential failure mode affects safe operation and/or involves noncompliance with regs with warning.	May endanger operator with warning.	9
Very High	Inoperable, loss of primary function.	100% of product may have to be scrapped.	8
High	Operable but at reduced level of performance; customer very dissatisfied.	Product may have to be sorted and a portion scrapped.	7
Moderate	Operable but convenience item(s) inoperable; customer dissatisfied.	A portion scrapped may have to be scrapped with no sorting.	6
Low	Operable but convenience item(s) operable at a reduced level of performance. Customer somewhat dissatisfied.	100% may have to be reworked.	5
Very Low	Fit and finish type items do not conform; defect noticed by most customers.	Product may have to be sorted with no scrap, and a portion reworked.	4
Minor	Fit and finish type items do not conform; defect noticed by about half of customers.	A portion of the product may have to be reworked out-of-station with no scrap.	3
Very Minor	Fit and finish type items do not conform; defect noticed by discriminating customers.	A portion of the product may have to be reworked in-station with no scrap.	2
None	No discernible effect.	Slight inconvenience.	1

Reprinted with permission from the **FMEA Manual** (DaimlerChrysler, Ford Motor Company, General Motors Supplier Quality Requirements Task Force).

Appendix 9
Standard (AIAG) Process FMEA
Occurrence Ranking Scale

(Should be tailored to meet the needs of your company.)

Description	Likely Failure Rate	P_{pk}	Ranking
Very High: **Persistent failures.**	≥ 100 per thousand pieces.	<0.55	**10**
	50 per thousand pieces.	≥0.55	**9**
High: Frequent **failures.**	20 per thousand pieces.	≥0.78	**8**
	10 per thousand pieces.	≥0.86	**7**
Moderate: **Occasional** **failures.**	5 per thousand pieces.	≥0.94	**6**
	2 per thousand pieces.	≥1.00	**5**
	1 per thousand pieces.	≥1.10	**4**
Low: Relatively **few failures.**	0.5 per thousand pieces.	≥1.20	**3**
	0.1 per thousand pieces.	≥1.30	**2**
Remote: Failure is **unlikely.**	≤0.010 per thousand pieces.	≥1.67	**1**

Reprinted with permission from the **FMEA Manual** (DaimlerChrysler, Ford Motor Company, General Motors Supplier Quality Requirements Task Force).

Appendix 10
Standard (AIAG) Process FMEA Detection Ranking Scale

(Should be tailored to meet the needs of your company.)

Rank	Detection	Criteria	Inspection Type			Suggested Range of Detection Methods
			A	B	C	
10	Almost Impossible	Absolute certainty of non-detection			X	Cannot detect or is not checked.
9	Very Remote	Controls will probably not detect			X	Control is achieved with indirect or random checks only.
8	Remote	Controls have poor chance of detection			X	Control is achieved with visual inspection only.
7	Very Low	Controls have poor chance of detection			X	Control is achieved with double visual inspection only.
6	Low	Controls may detect		X	X	Control is achieved with charting methods, such as Statistical Process Control (SPC).
5	Moderate	Controls may detect		X		Control is based on variable gauging after parts have left the station, or Go/NoGo gauging performed on 100% of the parts after parts have left the station.
4	Moderately High	Controls have good chance of detection	X	X		Error detection in subsequent operations, or gauging performed on set-up and first piece check (for set-up causes only).
3	High	Controls have good chance of detection	X	X		Error detection in-station or subsequent operations by multiple layers of acceptance: supply, select, install, verify. Cannot accept discrepant part.
2	Very High	Controls almost certain to detect	X	X		Error detection in-station (automatic gauging with automatic stop feature). Cannot pass discrepant part.
1	Very High	Controls certain to detect	X			Discrepant parts cannot be made because item has been error-proofed by process/product design.

Key for Inspection Type: A = Mistake-Proofed, B= Gauging, C = Manual Inspection

Reprinted with permission from the **FMEA Manual** (DaimlerChrysler, Ford Motor Company, General Motors Supplier Quality Requirements Task Force).

Appendix 11 (Page 1 of 5)
Examples of a Custom PFMEA Scales

Severity: PFMEA Operational Examples

Ranking	Example
10	Critical process equipment damaged and unusable or destroyed.
9	Loss of customer due to late delivery.
8	Entire lot of top-level assembly product scrapped.
7	Full assembly line (or bottleneck operation) down more than 1 week.
6	Rework full lot of top-level assemblies.
5	Scrap full lot of sub-level assemblies.
4	Technical (engineering) resources required to get line operational.
3	Rework sub-level assemblies off-line.
2	Equipment down for more than 1 hour.
1	Engineering disposition.

Severity: PFMEA Customer Satisfaction Examples

Ranking	Example
10	In-service failure that threatens safety.
9	Extensive product recall.
8	Unscheduled engine removal.
7	Premature (unscheduled) component replacement.
6	Oil leak but system still operational.
5	Air-conditioning system not operating properly.
4	Interior panel rattles.
3	Variation in seat colors.
2	Door plugs missing.
1	Scratch on interior of housing.

Appendix 11 (Page 2 of 5)

Severity: PFMEA EH&S Examples

Ranking	Example
10	Loss of life; serious injury.
9	Large hazardous material spill or release.
8	OSHA Recordable injury.
7	Personnel exposure above PEL.
6	Moderate hazardous material spill or release.
5	Fail internal ISO 14001 audit.
4	Injury requiring first aid.
3	Spill of non-hazardous material.
2	Minor (non-hazardous) coolant spill.
1	Poor housekeeping.

Occurrence: PFMEA Time-Based Examples

Ranking	Example
10	≥1 per occurrence per shift
9	≥1 per occurrence per day
8	≥1 per 2-3 days
7	≥1 per week
6	≥1 per 2 weeks
5	≥1 per month
4	≥1 per quarter
3	≥1 per half-year
2	≥1 per year
1	<1 per 1 year

Appendix 11 (Page 3 of 5)

Occurrence: PFMEA Piece-Based Examples

Ranking	Example
10	Cpk < 0.33
9	Cpk ≈ 0.33
8	Cpk ≈ 0.67
7	Cpk ≈ 0.83
6	Cpk ≈ 1.00
5	Cpk ≈ 1.17
4	Cpk ≈ 1.33
3	Cpk ≈ 1.67
2	Cpk ≈ 2.00
1	Cpk > 2.00

Occurrence: PFMEA Event-Based Occurrence Examples (or Examples for Complex Assemblies)

Ranking	Example
10	≥1:2 events (or complex assemblies)
9	≥1:10
8	≥1:25
7	≥1:50
6	≥1:100
5	≥1:500
4	≥1:1,000
3	≥1:5,000
2	≥1:10,000
1	<1:10,000

Appendix 11 (Page 4 of 5)

Detection (Control): PFMEA Mistake-Proofing Examples

Ranking	Example
10	
9	Does not apply.
8	
7	Sensory alert prevention solution; color-coding of drums of raw material.
6	Warning detection solution; audible alarm sounds if over-torque condition is detected with pump.
5	Warning prevention solution; alarm flashes if rate of pump motor torque rise is excessive.
4	Shutdown detection solution; pump shuts down if over-torque condition is detected.
3	Shutdown prevention solution; cycle counter with automated shutdown at MTTF (mean time to failure).
2	Forced control detection solution; automated in-line inspection fixture.
1	Forced control prevention solution; use of asymmetrical features to allow placement of fixture one and only one way.

Detection (Control): PFMEA Gauging Examples

Ranking	Example
10	Does not apply.
9	
8	Periodic NDT.
7	Periodic in-line variable gauging.
6	Periodic in-line GO/NOGO gauging.
5	In-line GO/NOGO gauge on all parts exiting process.
4	Automated inspection on first piece.
3	Dimensions of input materials confirmed with in-process accept/reject gauging.
2	100% automated inspection of 100% of product.
1	Does not apply.

Appendix 11 (Page 5 of 5)

Detection (Control): PFMEA Manual Detection Examples

Ranking	Example
10	No monitoring, measurement, or sampling.
9	AQL sampling plan used for Final Inspection.
8	100% visual inspection.
7	100% visual inspection with visual standards.
6	100% manually inspected using GO/NOGO gauges.
5	SPC used in-process with Cpk 1.33 or higher.
4	SPC used in-process with Cpk 1.67 or higher.
3	
2	Does not apply.
1	

Appendix 12
FMEA Analysis Worksheet

Process/Product: _____
FMEA Team: _____
Team Leader: _____

FMEA Process

Component & Function	Potential Failure Mode	Potential Effect(s) of Failure	Severity	Potential Cause(s) of Failure	Occurrence	Current Prevention Controls	Current Detection Controls	Detection	RPN	Recomm'd Action	Respons. & Target Completion Date

This form is shown over two pages to enhance readability. ➔

Appendix 12
Control Plan Attachment

FMEA Number: _____

FMEA Date: (Original) _____

(Revised) _____

Page _____ of _____

Action Results					Control Plan						
Action Taken	Severity	Occurrence	Detection	RPN	Control Factor	Spec or Tolerance	Measure Technique	Sample Size	Sample Freq	Control Method	Reaction Plan

← This form is shown over two pages to enhance readability.

Appendix 13
Glossary of Terms

AIAG	Automotive Industry Action Group, an organization with a mission to improve the global productivity of its members and the North American automotive industry.
Boundaries of Freedom	A conceptual management tool used to define and communicate predetermined levels of authority for the use of time, funds, and resources.
Bell-Shaped Curve	A pattern of variation known as the normal curve.
Bimodal Distribution	A pattern of variation which has two or more peaks, or modes.
Capable	If the process is stable, normally distributed, and the process spread (six standard deviations) is less than the customer's specification range (T.T.) with room to spare (industries today typically require 25%), the process is capable. Capable only means the process can fit within the specification. However, it may not fall within the specification. Ideally we want a process to be both capable and centered.
Capable and Centered	A process that is capable of meeting the specification and is operating in the approximate center of the specification; it has a Cp approximately equal to the Cpk and the Cpk is equal to or greater than 1.33.
Control Chart	A chart used to maintain statistical control of a process.
Control Plan	Written documentation of the systems to be used to control processes and/or product components.
C_{pk}	The C_{pk} is the best measure of process capability because it not only tells you if the process is capable, but also whether it is centered. C_{pk} = minimum of $\{C_{pu}, C_{pl}\}$. The C_{pu} measures the capability of the top half of the process and the C_{pl} measures the capability of the lower half of the process. The C_{pk} is like a bowling score - the higher the better. In order for a process to be considered capable, the C_{pk} should be at least 1.33.
Design of Experiments (DOE)	A family of statistically based techniques and methods used to conduct organized experimentation. DOE techniques require relatively little time and money compared to conventional experimental techniques. Yet, they can yield comprehensive information about the individual variables being studied as well as interactions between the variables.

Detection Ranking	A ranking scale ranging from 1 to 10. The easier it is to detect the cause of a failure or the subsequent failure, the better. A high ranking (e.g. 10) means the probability of detection is low. A low ranking (e.g. 1) means the probability of detection is certain.
DFA	The acronym for Design for Assembly. The purpose of DFA techniques is to design a product in a way that makes assembly easier for manufacturing. One example of a DFA technique would be redesigning a part that requires 10 screws in the assembly process so it could snap together instead thereby eliminating the need for any screws.
DFM	The acronym for Design for Manufacturability. The purpose of DFM techniques is to design a product in a way that makes it easier to manufacture. One example of a DFM technique is designing all products to use the same type of fastener rather than specifying a different type of fastener for each product.
DFMEA	The acronym for a Design-FMEA. A DFMEA is used to identify and evaluate the relative risks associated with a product design.
DOE	The acronym for design of experiments. (See Design of Experiments.)
Effect	The potential impact of a failure should the failure occur.
EVOP	The acronym for a type of DOE (see Design of Experiments) called "Evolutionary Operations." The EVOP experimental method targets an experimental region of variables within the current range of operating conditions of a process. Using this method, improvements in output and quality can be achieved without interrupting the process.
Failure Mode	How a product design or process component could fail and have a resulting impact on the product or process performance.
Flowchart	A graphical representation of a process sequence using standard symbols.
FMEA	The acronym for Failure Mode and Effects Analysis. An FMEA is a systematic, structured approach to identify and evaluate, on a relative scale, risks associated with a process or product.
Fractional Factorial	A derivative of the full factorial DOE (see Design of Experiments) in which higher order interactions are assumed unimportant. This reduces the total number of experimental runs to a "fraction" of the number which would be required for a full factorial.

Frequency	The number of times a value or event occurs.
Full Factorial	A type of DOE (see Design of Experiments) in which every possible combination of factors and levels are studied so that the main factor effects and all interactions can be studied.
GR&R Study	The acronym for Gage Repeatability and Reproducibility studies. A GR&R is a study of the variation in the measurement system.
In-Specification	A process or data point from a process that falls within the specification limits.
LCL	The lower control limit of a control chart.
LSL	The lower limit of a specification.
Mean	The arithmetic average for a group of values. Also known as the $\overline{X}$ (x-bar).
Mistake-Proofing	A technique that makes a process or product so robust that it cannot fail. Also known as poka yoke or error-proofing
Mixture Experiment	A special type of DOE (see Design of Experiments) used when studying the effect of varying proportions of components of a mixture. Mixture experiment techniques assure that the sum of the components always adds up to 100% and that each component is accounted for in the experimental design.
Nominal	The value the customer ideally wants for a product parameter.
Normal Distribution	A pattern of variation of a stable process (one with no special causes) in which the distribution resembles a bell-shaped curve.
Occurrence Ranking	A ranking scale, from 1 to 10, used to evaluate how frequently a failure due to a specific cause will occur. A high ranking (e.g. 10) means the failure due to the given cause results occurs very often. A low ranking (e.g. 1) means the failure due to the given cause rarely or never occurs.
Out-of-Specification	A process or data point from a process that falls outside of the specification.
Out-of-Control	A process that is not stable due to special causes of variation.

Pareto Principle	The Pareto Principle is often called the 80/20 rule. It means that approximately 20% of categories account for approximately 80% of the total impact.
PFMEA	The acronym for a Process-FMEA. A PFMEA is used to identify and evaluate the relative risks associated with a process.
Pilot Process	A pilot process is a small-scale operation used to develop new products or to test out process modifications before moving it up to production operations.
Poka Yoke	A technique that makes a process or product so robust that it cannot fail. Also known as mistake-proofing.
PPM	Parts per million.
Process Capability	A measure used to determine if a process output is capable of meeting its specification.
Prototype	A prototype is an early version of a new product. Often the prototype is not a working model or may even be a mock-up.
R	Range. A measure of the dispersion or variation in data.
Range	Also known as R. The highest value in a data set minus the lowest value in the same group. Describes the dispersion in our data.
Reaction Plan	The Reaction Plan is part of the Control Plan. It spells out how the organization will react if a failure mode occurs.
Risk	Risk refers to a potential hazard or quality deficiency associated with a product or process.
RPN	The acronym for Risk Priority Number. The RPN is a product of multiplying the Severity ranking x the Occurrence ranking x the Detection ranking. The RPN will always be a number from 1 to 1000, where 1000 is the maximum risk.
s	The standard deviation of a sample of data.
Sample	A representative subset of data randomly taken from a population of data.
Screening Experiment	A screening experiment is a type of DOE (see Design of Experiments). It is a severe fractional factorial that allows many factors to be studied with relatively few experimental runs.

Severity Ranking	A ranking scale, from 1 to 10, used to evaluate the relative severity of the consequence of a failure. A high ranking (i.e. 10) means the consequence or impact is grave. A low ranking (i.e. 1) indicates that the impact is minimal or unnoticeable.
Six Sigma Quality	Theoretically, a process with a Cpk of 2.0 and 2 ppb defects. However, when used to describe "Six-Sigma" as used by many companies today for "Six-Sigma Quality," it refers to 3.4 ppm quality and not 2 ppb. The reason for this difference is that the Six-Sigma community accounts for long-term process drift that some statisticians have estimated to be approximately 1.5s. Thus a distribution that has ± 6s within the specification and then drifts 1.5s actually has its mean at 4.5s from one of the specification limits at times. Looking only at that tail of the normal curve as being outside the specification gives us the 3.4 ppm (½ of 6.8 ppm in a 4 .5s) quality level.
Skewed Distribution	A pattern of variation which is non-normal; it appears "pushed over" to one side.
SPC	The acronym for Statistical Process Control.
Special Cause of Variation	A cause of variation which is unpredictable and makes the process unstable.
Stable Process	A process that is in-control with only common causes of variation present.
Standard Deviation	A calculation on a set of data that indicates how much variation there is in the data.
Subgroup	A small grouping of samples. For control charts, subgroup sizes usually range from 2 to 5 depending on the variation in the process.
UCL	The upper control limit on a control chart.
USL	The upper limit of a specification.
Variable Data	Data that are measured on a continuous scale.
Variation	The difference between similar items or things.
X	An individual data point or observation.

X-bar	The arithmetic average for a group of values. Also known as the mean.
X-axis	The horizontal axis on a graph.
Y-axis	The vertical axis on a graph.

Using your Training Program

System Requirements

✎ This program will run on most Pentium PCs.

✎ Information on system requirements can be found at: www.qualitytrainingportal.com/support.

License

✎ CD-ROM Version

● The program is licensed to be used on one computer.
● See the **cdlicn.txt** file on the CD-ROM for license information on the CD-ROM version.

✎ LAN Version

● The LAN version is licensed by the number of concurrent users at one physical site.
● See the **lanlicn.txt** file file on the CD-ROM for license information on the LAN version.

✎ WBT Version

● Each learner is assigned a unique username with a corresponding password. One and only one learner is licensed to access the WBT program using the username/password combination.

Installation

✎ CD-ROM Version

● See the **cdinstall.txt** file on the CD-ROM for installation instructions or go to www.qualitytrainingportal.com/support.

✎ LAN Version

● See the **laninstall.txt** file on the CD-ROM for installation instructions or go to www.qualitytrainingportal.com/support.

✎ WBT Version

● No installation is necessary. WBT learners access the program from a computer connected to the internet using their log-in information.

● IE 5.0 or higher is required. Java must be enabled.

Technical Support

Using the *FMEA Investigator* should be easy, but if you experience any problems, we're here to help. Before you call our technical support line, please check the following:

✎ Confirm that your system has the minimum requirements necessary to run the CBT program. If not, you will need to upgrade your equipment.

✎ Shut down your system and restart it again. Sometimes this can fix the problem.

✎ Go to www.qualitytrainingportal.com/support to review known technical problems and fixes to those problems. (This address is case sensitive; please use all lower case letters.)

● If you are still experiencing problems, try to determine if it is a hardware problem before contacting us. We will not be able to help you with hardware issues.

● If you have gone through the above steps and are still having problems, e-mail us at support@reseng.com or call our technical support line at (888) 810-8326 or (860) 872-2100.

WBT vs. LAN vs. CD-ROM Delivery Formats

Resource Engineering offers all programs in three delivery formats: WBT (web-based training), LAN (local-area network), and CD-ROM. The best option for an organization depends on their objectives and structure. The advantages of each option are recapped below:

✎ Web-Based Training (WBT)

- Access anytime, anywhere.
- Unlimited concurrent users.
- Pay by use.
- Access all courses without a large initial investment.
- Centralized training records across many sites.

✎ LAN-Based Training

- Use on any LAN computer at the physical site.
- Train multiple people with additional concurrent users.
- Fixed investment with optional maintenance and support.
- Centralized training records for the site.

✎ CD-ROM-Based Training

- Lowest fixed cost investment with optional maintenance and support.
- Train unlimited employees.
- The more you train, the lower your training costs.

In summary:

- **WBT**
  - ✓ The WBT approach is the most effective approach for organizations with many locations providing learners with training on a variety of subjects.
- **LAN**
  - ✓ A LAN solution is best suited for organization with many people to train on one program (or a limited number of programs) at one site.
- **CD-ROM**
  - ✓ A CD-ROM is the best option if an organization has a limited number of learners to train on a limited number of subjects.
  - ✓ CD-ROM delivery may also be an effective approach for organizations that use a training center with dedicated computers for training.
- **All approaches:**
  - ✓ Provide documentation of training.
  - ✓ Include competency testing.
  - ✓ Build a database of learner's performance against the tests.

Other CBT & WBT Programs

Resource Engineering is constantly developing new CBT and WBT programs to support quality and productivity improvement initiatives. Some of the programs available are:

✎ Mistake-Proof It!
- Applying poka yoke, or error-proofing, techniques.

✎ SPC Workout
- Basic SPC including measures of variation, using control charts, and measures of process capability.

✎ Advanced SPC
- Setting up control charts, conducting process capability studies, and understanding Six Sigma capability.

✎ Measurement System Analysis
- Analyzing measurement system variation (GR&R studies) and managing measurement systems.

✎ Screening Experiments
- Design of Experiments (DOE) techniques focusing on Taguchi and Plackett-Burman methods.

✎ Gage Mentor
- Use and care of dimensional gages.

✎ Problem-Solving Techniques
- Problem-solving tools & techniques using the 8-Discipline approach.

✎ Six Sigma Start-Up
- An overview of the principles & practices of Six Sigma.

✎ DMAIC Tools and Techniques
- Problem-solving tools & techniques using the Six Sigma DMAIC approach.

✎ Lean Manufacturing
- An overview of the principles & practices of Lean Manufacturing.

Check out www.qualitytrainingportal.com to see what new programs are available!